"CREATED FOR HIS GLORY"

LIFE INTIMATE CONVERSATIONS WITH THE FATHER

Angela Lewis

Contents

Dedication	i
Acknowledgements	ii
Prologue	iii
Chapter 1 Face To Face	1
Hungry Love	2
All Eyes On Him	6
Hope	8
Father	10
Chapter 2 Experiencing Jesus	12
Yeshua	15
Taste and See	16
Comforter	17
Glorify Him	18
Chapter 3 The Crucible Of Life	19
Take Another Step	20
Beauty Within	22
Healing Journey	24
Freedom Song	26
Better	27
Connect	29

Chapter 4 Joy Speaks — 30
- Lighthouse — 31
- My Word — 32
- Creator-Create — 34
- To Be Known — 36
- Heavenly Connection — 38
- Thy Word — 40

Chapter 5 I Am He — 41
- Bread of Life — 42
- Eternal — 43
- Healing — 44
- Secrets — 45

Chapter 6 Come To Me — 46
- Midnight Cry — 47
- Let's Talk — 49
- Walk With Me — 50
- Eagles Wings — 51
- I'm the Answer — 52
- Jesus — 53
- Bleeding Heart — 54

Chapter 7 A New Beginning — 55
- Great Delight — 56
- The Time is Right — 57
- Captured — 58

Limitless	59
Rise Above	60
Instructions	61
Chapter 8 Time To Soar	**62**
Extraordinary	63
Guided By The Light	65
Spiritual Formation	66
Rest In Me	68
Triumph	70
Citations	**71**
References	**72**

Dedication

This book of intimate reflections is dedicated to the Father for His captivating love that He has given to His daughter. Throughout life's precious moments (Daddy Pa Pa) has provided His love, comfort and reassurance.

It is my hope that all that read these penned lines will sense the genuineness of the Father's Love and His loving arms which await them. The Father is waiting for you to receive His indescribable love. He will walk through the dead places of your life with you. You are never alone.

Daddy, thank you for loving me/us relentlessly.

Your Daughter

***Angela Lewis**al*

Acknowledgements

It is my deepest desire to acknowledge those who have been a pivotal part of my journey.

To my loving husband (Jeffrey Lewis), thank you for your steadfast love and unwavering commitment to God and to me as we have vowed to live for Jesus. Through our successes and failures, God has remained ever so faithful and true.

To my family, my loving mother Geneva Fuller, who has always believed the best for me and whose strong determination and grit to love God with all her heart and invest it into her children, my soul truly thanks you, my dearest momma. To my step-father, Mose Fuller thank you for supportively raising my sister and I.

To my beloved sister, Carla McGee, a giver at the heart of her time, talent, and treasure. Thank you for always challenging me to grow and win. Thank you for your love and everything you do for so many people.

To my loving family, thank you for being who you are; I love you.

Thank You to my wonderful friends and heartfelt soulful sisters Diane, Denise, Patrice, and Francis, and Chaplain Sisters Diana, Barbara, and Janaclese. Thank you to my wonderful pastors, Michael, Chris, and Jonathan Fletcher, Ron Butler, Ben Goodman, and David Teague, Bishop Porter, elders, and leaders who have invested the unadulterated Word of God, love, and mentorship to help me grow and strengthen me throughout the vicissitudes of life. Thank you for doing life with me and ensuring that I would be willing to face the wounds, allow God access to heal, and birth out all that God has poured into me.

In closing to my best friend who has passed on into Glory (Faith) Rosalyn Liboy, Oh, how I miss you so dearly! My life has forever been changed by encountering you as a Sistah friend. I miss you dearly, and I'll love you forever.

Prologue

The purpose of this book is to encourage and empower readers to take the time to reflect on their lives and the conversations they have with the Father. These intimate, poetic conversations with the Father provide strength, guidance, comfort, and hope to face the future. These conversations provide an opportunity to tap into what is causing disease within the soul as the Father provides healing.

I have found so much freedom within the arms of the Father as He has wrapped me in His arms of love and assurance. Above all else, these intimate conversations have enriched my relationship and drew me closer to PaPa. He is the ultimate healer in whom I take refuge.

Authenticity and transparency are available to you today! Are you ready to take the plunge and be all you have been created to be? The Time Is Now!!!

Chapter 1
Face To Face

Hungry Love

Blessed are those who hunger and thirst for righteousness, for they shall be filled

(Matthew 5:6)

I have been graced from above
And you have filled me with your hungry love.
Oh, what a sweet delight to find such favor in your sight.
I hunger and thirst after you alone
as forces compete for my undivided attention to be drawn.
You alone have I searched for all my life
Although many substitutes have engaged my plight.

Nevertheless, you alone have brought me the ultimate peace
and exchanged the beauty for ashes that showed up within me.
Substitutes will never suffice
because only you can satisfy the ache within my soul.
It's your love alone that touches me and makes me whole.
Draw me into your warm embrace
as I bask in the goodness of your sweet embrace.
Refresh my soul again for
in you I always win.

Just to behold you and call you mine
I know I am accepted and beloved
by your matchless love alone.
So hungry for the love I can call my own.
I can't get enough of you (ABBA)....

ABBA

Though my father and mother forsake me, the LORD will receive me (Psalm 27:10)

I call you (Father), Daddy, Papa

because you make yourself known to me

In all your majesty, you allow me to know thee.

How is this that you have a love only for me that reaches far beyond what my eyes can see?

I can't even begin to comprehend this love that I have been given like I've never known.

Your loving lessons have taught me what it means to be fully secure, loved and known.

No longer hiding and waiting to be known, I run full speed into your loving arms.

I am free to be authentically me

because of your love and truth that flows through me.

I am never rejected or left alone.

You have taught me what it means to be truly Fathered in your love alone.

A love so pure that it can stand on its own.

Thank you for taking on the sins of this world and everything that I have done.

Thank you for going the distance and experiencing the pain of rejection.

Thank you for taking it on for me and everyone before and after me.

I can't even begin to define or describe your love in mere words alone, but it is embedded in my heart and entrenched within my soul. You have captured my heart.

I call you (FATHER)

All Eyes On Him

Fixing our eyes on Jesus, the pioneer and perfector of faith.
(Hebrews 12:2)

Set your gaze as you arise from the haze.
Your moment awaits to give Him your perpetual and earnest praise.
Do not stuff it down. It's time to blast this majestic sound.
Let the world know all around
that He is the eternal and almighty one
Worthy of all Glory and
of which no one can compare.

Can't you see Him?
Eyes A blazing like the Sun
I arise to sing to the Holy One!
Daddy, I'm here.
You have captured my attention.
You alone take me deeper into incredible dimensions.

Your love is so sweet.
You are so unique.
Thank you for allowing me to get to know thee.
All I want to do is show you how much I love you.

You are loved.

It's like I've been kissed from above.

My eyes are on you.

You hold all my attention.

Hope

Blessed be the God and Father of our Lord Jesus Christ, who according to His abundant mercy has caused us to be born again to a living hope through the resurrection of Jesus Christ from the dead (1 Peter 1:3)

You are the hope that I have experienced like I have never known.

You promised me that you would never leave me alone.

I can trust and believe in what your words say that

you will never leave me or forsake me!

Each new day that the sun arises.

Hope begins again, you are my one true friend.

A new day's dawning helps me to breath in the breath of life again

New Hope Begins

Each day I am commanded to trust in you and not just what I see.

You have truly got a heavenly grip on me.

This is where true hope and faith awakens me.

Sometimes it so hard not to go my own way.

The tug at my heart is forever beckoning me to sit in your presence to reflect on the price you've

PAID.....

Daddy Show me and teach me your righteous ways, Yahweh.

Guide my path and keep me from going astray.

I am refreshed within my soul as hope awakens.

Once again, I find strength beyond what I can dare try to explain.

You have broken off the chains on my brain.

You transformed my mind once again.

What hope I have found in my dearest friend.

Hope Restored!

Father

And I will be a father to you, and you shall be sons and daughters to me, says the Lord Almighty. (2 Corinthians 6:18)

Of all the greatest gifts I have ever known, you gave me life!

You drew me to yourself and called me your own.

Knowing that I would never be the same.

You knew me before I knew myself.

Yet you rode the wave with me to figure out my zig zag quandary.

And you never left me alone.

Always standing by with open wide arms and Grace filled eyes....

I need you for the rest of my life

You are the Ancient of days...

I have never known the love of a biological Father... yet and still you showed me what it means to be Fathered by the Matchless One....

(Daddy Papa) I adore thee....

Oh, how I want to become selfish with your love

You let me know you have so much love to go around, and you are concerned about every trivial detail of my life.

So many times, I have cried in your arms as you have held me up when I was falling down.

You called me to yourself and embraced me in a love that is perpetual.

Daddy's Daughter

To Know You and Make You Known

You shall love the Lord your God with all your heart and with all your soul and with all of your strength and with all your mind, and your neighbor as yourself. (Luke 10:27)

It is so beautiful to get to know you.

Bread of Life, for you have called me your own.

You are the Living Word; I behold the true essence of your face.

A face like one I've never known, and suddenly, my heart begins to yearn.

Hungry and thirsty, I thirst for you more

for my aching heart attests that there is so much more

Just to know you are not enough

I must share about your KINGDOM that HAS COME

With those who do not know the unforgettable one

You love them with a love so dear

When they witness it, they will want to draw near

Heal them from the pain of their past

So, they will be able to say and see at last

No one could ever love them with such grandeur allure

You are our everything for sure.

Chapter 2
Experiencing Jesus

I'm on my Way

But I have this against you that you have abandoned the love you had at first. (Revelation 2:4)

I hear you beckoning me to come.

I am almost at full stride as I realized I was running in the wrong direction.

You captured me running away

And derailed the path that led me astray

Now I have you in my sight as I set out for the greatest marathon of my life

I'm running full speed ahead straight into my Master's Arms.

I have been drawn to sit at My Master's Feet

To hear His voice that captures me

My heart is thirsty for the eternal water that fills

My dessert soul

When I drink of the eternal water that fills my soul, my heart is set ablaze

I won't allow the world to extinguish its fiery blaze.

Your flame burns bright and deep within my soul.

It is an unexplainable feeling that I can't control.

I know it has been a while since I have been in this place, and I want to

Thank you (Daddy, Pa Pa) for receiving me with open arms.

You are My Savior, My Lord, My Redeemer
I am here to say I'm thankful to be home
Back in my rightful place bowing before for my master's face.
Thank you for choosing to wait for me.
I am Yours my Beloved!

Yeshua

For the Son of Man came to seek and to save the lost (Luke 19:10)

You are indeed my rescuer, deliver, my hope, my friend, and my Lord. Love

cannot be articulated without starting with you as the reference point. I start and finish with you.

Unnumbered and countless times you have rescued me from myself! From the thoughts that raced within my mind dashing about and trying to understanding life's challenges of mind.

You demonstrate what it means to be rescued and delivered from the doldrums of life while

Focusing all of my attention on the one that has given me life.

You are my living hope and the reason why I have breath to breathe

My Dear Yeshua

You are Lord of All and my friend

I find my solace and significance in you alone

I rest in your peace Shalom

To be with you is where I'll spend my eternal home.

Taste and See

Oh how I love your law! It is my meditation all the day (Psalm 119:97)

You are the living Word that nourishes my parched soul. I drink of you this day and forever more. Your goodness leaves me in awe of your majesty and your power.

I want to know you more and more.... you make my heart desire to explore all of that you have in store. What beauty I see and behold every day.

I grab a hold of the richness of your Word that leaves such a delicious delight.

May I never shrink back from your sight.

Oh, beautiful one! I desire you.

I open my eyes and behold the goodness of your creation

I look around and know that there is

ONLY ONE GOD

Whom deserves all honor and glory.

I breathe you in one thousand times again and again

How could I ever get enough of your LOVE

I have been captured and cultivated by your sweet and enduring love.

Comforter

However when He, the Spirit of Truth, is come, He will guide you into all truth; for He shall not speak from Himself, but whatsoever He shall hear, that shall He speak; and He will show you things to come. (John 16:13)

Teach me, Lead Me, Guide me in the way I shall go. I will follow even when I do not know.

You bring forth truth like I have never known, and you go to the waste places, to give me beauty for the ashes, oh what peace and serenity I find in you. You know me completely through and through. The wisdom you give me comes from above and flourishes and illuminate through me the riches of your Devine Love.

Glorify Him

***For everything comes from him and exists by his power and is intended for his glory. All glory to him forever! Amen
(Romans 11:36)***

Let me only seek to Glorify the majestic one that His name may be glorified in the earth. As I arise each and every day, my greatest desire is to bring you the praise and glory that you deserve.

You are to be glorified above them all, for no one can match your goodness and your love

I honor you "Father" for all that you represent.

You are my EVERYTHING!!!!

Take your rightful place in all that you have created because you have allowed us to experience life.

You are my life and without you in it, my breath is not worth breathing

What would it profit me to have everything, and without you, I have nothing.

You are magnified in the highest all Glory and honor to you alone.

Chapter 3
The Crucible Of Life

Take Another Step

I will instruct you and teach you in the way you should go: I will counsel you with my eye upon you. (Psalm 32:8)

Don't move right

Don't move left

There are many more steps to accomplish your goal

many voices calling you away from your assigned path

stay in tune

with the voice of the creator

listen well and be still your soul

He will lead you and guide you

Right where you need to be

Yes, there may be many times you are

Ready to flee

Trust His plan and take heed

For when you take His hand

There is a rhythmic grace within the Master's Plan

Look up, stay focus, ad fix your gaze

For He is making your path clearer along the way

Even when you fill like you're operating within a maze

He knows your assignment comprehensively.

As you walk, he is revealing divine insight along the way

Revelation will come as you pass by His ordered and directed way

You will arrive at your pre-destined place

If only you allow him to mature you along the way

He will cut you, gut you of all that is not needed along the way

To remove the clutter and non-essential things that are trying to get in your way

He will make you moldable in the Master's hands

And you will arrive on time according to His Master Plan

He has been your GPS every step of the way

Never missing a turn but guiding you safely along the straight and narrow pathway

Right to the place, where he has desired for you since the beginning of time

Now it is your time to soar, are you ready?

Beauty Within

To appoint unto them that mourn in Zion, to give unto them beauty for ashes, the oil of joy for mourning. The garment of praise for the spirit of heaviness. (Isaiah 61:3)

Who are you, my daughter?

You are chosen

Can you see the beauty within

You are not like any other

For there is no replica that could ever suffice for you.

For I paid the price with all my life.

Let me tell you who you are so you will never forget.

You are one of a kind, special and unique, created distinctively by me.

No matter how many times life screams at you about who you are not

You are my daughter

One filled with beauty inside and out

One whom I have called my own

That I will never leave alone

A woman of strength, power, meekness, and grace

Poetic in a sense that leaves her trace

Handcrafted by the master architect, chief builder, and cornerstone

You ar valued and loved beyond your greatest imagination.

Remember my Words of who you are

Reflect me, in Christlike Character, love, joy, hope, and truth.

You are mine

Healing Journey

For I am the LORD that health thee. (Exodus 15:26)

Along this journey there has been many hard roads along the way

I must say, some that I have not desired to cross along the way

Nevertheless, along all, I've heard your voice continually call Me to yourself.

I failed in life and have been restored.

I've been challenged to look beyond the failures to the lessons that have called me higher

I've been called to face the unknowns of sickness, wounds of the soul, heartbreak and pain

Sometimes, graveling in the pain

But through it all, I have been able to bounce back

It is you, Lord that has built me with strong resiliency after the fall

Now I can relish in it all that you have wasted nothing along this journey

That has been surrendered to you

You have used it and still use it all for your Glory, indeed

Now, I am forced to take heed

Thank you for the wounds that caused me to triumph and reach for you

I will forever keep reaching for you with all of my heart

Yes, I have experienced how life has tried to rip me all apart

I'm still here not because of my own strength but your strength has been perfect in my weakness

I arise on this healing journey to say to another dear soul along the way

Never give up or in along the way

There is true healing in the master that will never leave you or throw you away.

Freedom Song

It is for freedom that Christ has set us free. Stand firm, then, and do not let yourselves be burdened again by a yoke of slavery. (Galatians 5:1)

In you I find the freedom to be
Only who you created me to be
In you alone is where my life sings the most beautiful song
I stand amazed at how free I can be
When I entrust myself to the matchless one

I'm running and dancing around so free
Without a care of the world taking a hold of me
Daddy, Do you see me?
I am free to be just you and me

My song to thee I sing with my life
May the aroma be breathtaking
Without a hint of strife
As I draw near to the one that has called me His own
This is my beautiful heart song
You have written it all over me
And I want the world to see the love you put within me

Better

Better is a little with the fear of the LORD, Than great treasure with trouble. (Proverbs 15:18)

I'm a better me because you have

Touched the innermost part of me

With you there is no pretending

I stand in your presence flawed and all

You come to my rescue, and I hear your divine call

I'm better because you chose me

It is nothing I could dare to do or become that would make you prouder of me

I have been accepted and I am loved

I do not have to strive for perfection for your love

You sanctify me and cleanse me from the weight of this world

I come bold and truthful before your throne of Grace and Love

I will not hide myself from thee

You already know the REAL ME!

I behold you and I want to be better each and everyday

You draw me into your love each day

In each and everyway

I mature in your presence
So may I stay knelt at your feet
Your love keeps captivating me!

Connect

But seek first the Kingdom of God and His righteousness, and all these things shall be added to you. (Matthew 6:33)

How is it that I connect so well with thee.

My heart pours out the intricate details of my souls

Tears stream down my face in release of the agony and pain

All shame is washed away in your presence alone

I keep coming back to the foot of the throne

I sense the warm embrace upon my shoulders ever so light

And the comfort that fills my soul ever so tight

It is just what I needed when I never knew that is what I was searching for

I no longer want to hide from you anymore

Not as if I ever could....

However, I am free to connect with the all-powerful, omnipresent, and omniscient God

Of the UNIVERSE

Draw me near to your loving arms

That has wrapped me tight through so many dark nights

I choose to chase after you

There is no other connection greater than the one you have allowed me to have

I am YOURS!

Chapter 4
Joy Speaks

Lighthouse

Jesus said to him, "I am the way, and the truth, and the life. No one comes to the Father except through me. (John 14:6)

When all of life seems dark and you are looking for answers near and far

Turn toward me and look up so that you can see

My light is bright enough for the whole world to see

But all to often many look away from me

They want the answers in record giving time

But never look to the one who keeps it all in line

You want to see through the darkness and pain of your life

Look to me and you will find me

I light the dark paths along the way

I keep them all from going astray

I am the one that knows the way

I will never lead you astray

Call to me and you will find your answers

Just be ready to receive what I must give

This will be the answer and the way you are to live.

My LIGHT shines BRIGHT!!!!

My Word

In the beginning was the Word, and the Word was with God, and the Word was God.

(John 1:1)

What is your Word Lord?
My Word is Power
It gives light to the darkest of hours
It is hope where there is nothing but pain
It cuts through the darkness and eliminates the pain
It washes away the shame
It cuts to the chase and obliterates the games

It is the water to the dead and dying soul
It transforms the ugliest soul
My Word is hope that shines even with a glimmer of light
It shines when one is blind and has no sight

It is absolute truth
And it does not change
When others have called it foolish it
Still upholds under the shame

Rich and secure are all those that delve in

They will experience the beauty of Word

Transforming From Within

My Word You Ask

Tell Me How It Has Transformed Thee

Creator-Create

Do you not know? Have you not heard? The Everlasting God, the LORD the Creator of the ends of the earth-does not become weary or tired. His understanding is inscrutable.

(Isaiah 40:28)

The power of Life and Death is What I have Given
I spoke and there was
If I am the creator that creates all things, and I AM
What have I given you to speak into existence

You will use your mouth for Death or Life
You have a Choice of what you desire to see in your life
So crying about what you do not have

Speak out and walk in all that you do have
I have given you the Keys to the Kingdom
What are you doing with all you have been given

Life I short and time is running out
Make your life count for the good
You know that you have everything that you should
I have spoken into you and told you how fearfully and wonderfully made you are...
Do you believe what I have said that you are?

Go away now and begin to speak....New Life....Has Begun

Don't Waste Your Life or Your Tongue

Speak Life

To Be Known

My Father has entrusted everything to me. No one truly knows the Son except the father, and no one truly knows the Father except the Son and those to whom the Son chooses to reveal him. (Matthew 11:27)

You allow me to be known! To be free to come out of my comfort zone!

Free to roam without any hesitation or reservation!

I want to know you in the fullness of all you are....

Meet with you Face to Face as you allowed your

Servant Moses to meet

I desire nothing greater than to sit at your Holy Feet

To feast on Living Word that captures and cultivates me

Time stands still as I sit and embrace and look up toward your face

Your presence is one like I have never known

It has taken my breath away time and time again

Lord, you have been tried and tested time and time again

Absolute Truth, Eternal Water, Bread of Life, Refuge, Healer, Savior,

What else do I need for you have encompassed it all within thee.

To know you allows me to know myself and others

The Greatest Gift I could ever receive

Thank you for taking the time to know me.

Daddy Abba

Heavenly Connection

***I will instruct thee and teach thee in the way which thou shalt go:
I will guide thee with mine eye (Psalm 32:8)***

The more I come into agreement with who I am in you
The deeper my heavenly connection becomes ingrained
You are my everything and knowing you
Drives me deeper in praise, worship, and the truth of understanding
That this world is not my home......

I live in this world but I'm not of this world!
May I never allow the weight of this world to quinch
My Heavenly connection with me and my Daddy Pa Pa.
I stretch forth to receive all you have declared of me

I reach beyond what is merely seeable with my natural sight
I call forth by the power of life and death
All that you have already predestined and ordained
To come forth
I speak life with the power and authority that you have given to me
It is not my own words, but it is the power that was written in the
Book of My Lord Adonai
I connect with the King of Kings, The Matchless, King of Glory

You are my life, and my Life is hidden in you.
I arise in the newness of all you have created and spoken of me
To live for you and die for you is gain
While I'm still here may my life bring you glory
And testify of you as the light of the world.

Thy Word

For ever; O Lord, thy word is settled in heaven (Psalms 119:89)

Your word IS LIFE

IT'S Sharper than any knife

It has cut through my life

Opened my eyes to what is right

With it I rise above the vicissitudes of life

Standing bold & strong

Taking my place where I belong

They Word has been sealed within my soul

Ready to release and encourage the most parched soul

Thy Word fills me up to the fill

Chapter 5
I Am He

Bread of Life

And Jesus said unto them, I am the bread of life: he that cometh to me shall never hunger; and he that believeth on me shall never thirst. (John 6:35)

I eat of your word and you show me how to live
You are my life and I do not follow in my own way
You are my compass, road map, and GPS that
Guides me every step of the way as
I live day by day

You guide me to the exact destination
And I am never misaligned
When I follow the path that you have designed
And move patiently within your set time

As I apply what you say
I receive strength for another day.

Eternal

Unto thee it was shewed, that thou mightiest know that the LORD he is God; there is none else beside him. (Deuteronomy 4:35)

Who Is the only one eternal one
The one that was and is and will be forever
The one that is called by "The Great I Am"
He is the Everlasting to Everlasting
King of Glory

There is no one that can compare
To the magnitude of His Glory
His Life validates the Creation Story

He is the Self-Proclaimed God of all creation
The Alpha and Omega
He needs no introduction
Creation bears His very existence

He Is Eternal God

Healing

He health the broken in heart, and bindeth up their wounds.
(Psalm 147:3)

I experienced the greatest healing anyone could imagine
I have been touched by the Master's Mighty Hand
He's held me close and touched the inside of every wound

The surgery was one needed for far too long
Though at times I ran away from His healing arms
Trying to fix the pain that was buried so deep within

Realizing time and time again
I was no match for what lied within
Learning daily how to relinquish control

I bent my knees and cried out for the Master to take control
The surrender was painful and needful just the same
In His loving arms I release all my pain and shame

His healing has won me over
His love has cleansed my wayward heart
I am SAFE!!!!

Secrets

He answered and said unto them, because it is given unto you to know the mysteries of the kingdom of heaven, but to them it is not given (Matthew 13:11).

Can be deadly and cause so much pain

But there is a Master who has a secret we all could accept and proclaim

His name is Jesus and He gave His life for all

His love is the greatest inheritance of them all

It will fill you up when you are at your wit's end

Stand the test of time far greater than your best friend

He is the secret that I cannot keep to myself

His love has captured my heart and His story I must tell

When you see me smiling, just know His love kept me from a burning hell

I want to share this secret with you indeed

Never forced but meet you at your greatest point of need

I tapped into Him and that is what He did for me

Now I'm willing to share His secret with all that are willing to hear me...

Chapter 6
Come To Me

Midnight Cry

Thou tellest my wanderings: put thou my tears into thy bottle: are they not in thy book.

(Psalm 56:8)

Have you ever cried in the night when all was silent in your bed while resting your weary head?

Where your tears streaming down your aching face?

While clinching your pillow ever so close to your tear-mangled face

Did you try to drown out the noise of the sobs and the pains that raced deep within your heart

I know this type of pain.....

Away from the world so no one can see, in the comfort of my home sometimes hiding away from me

Screaming out through tears for where no words could express

How do I make sense of all this mess

I've cried myself to sleep many days in the dark

With the weight of the world on my weary heart

Looking for solace and for someone to understand the pain

It's been hard to articulate, and I've been so drained

Daddy, you never left me alone in the rough, rugged, and raw
You reached down and captured me with your first draw...

Help me make sense of it all
Show me who I am and I will RISE ABOVE IT ALL.

Let's Talk

Come unto me, all ye that labour and are heavy laden, and I will give you. (Matthew 11:28)

Daddy Pa Pa, You have told me that I can come to you and tell you what's on my mind

Thing is sometimes I really don't know where to begin

You know me all so well and my speech may take you on ride that may never end

Who am I kidding you know me from the front, middle, and the end

I still have questions that I ponder in my heart

Help me to get them out and share all with you

There is no judgment when I pour out my complaint to you

You listen so freely as I go on and on about my life

Help me to stop and listen to what you have to say

Your Word is healing to my aching soul

Help thy servant to trust that you have all the answers to what I have to say

I can put my hand in your hand and trust you will navigate my life with perfection every step of the way

Walk With Me

Jesus answered, "I am the way and the truth and the life. No one comes to the Father except through me (John 14:6)

Walk with me and listen to what I have to say

I promise you I will not lead you in the wrong way.

Walk with me and close your eyes

I promise you will never fall while I'm guiding your way

Walk with me toward growth and maturity which I am calling forth out of you

Leave the path that is calling you to fret and try to figure things out on your own

I am hear for you and I know the way

I am the Way, Truth, and Life

Eagles Wings

"But they that wait upon the Lord shall renew their strength; they shall mount up on with wings like eagles; they shall run and not be weary; they shall walk and not faint".
(Isaiah 40:31)

You say you want to soar like and eagle and fly above all things

But there are too many weights holding you down and keeping you grounded and still

Release the chains that have claimed your brain

For there is victory in my mighty name

Don't allow the weights of this world to keep you beached to the ground

Run to me and find the freedom that your soul seeks

I assure you I have the answer that your soul seeks

Soar beyond what has stifled you at your core

You were meant to fly and soar

It does not matter what naysayers have to say

Claim this day as the victory that you dare to see

Spread your wings wide and soar above what is beneath

Set your gaze and hold it in your sight

You will soar high far beyond your once limited sight

I'm the Answer

I love those who love me, and those who seek me early shall find me. (Proverbs 8:17)

*Often you are looking for answers to life problems,
but you look to everyone but me*

*Why do you choose to look for the temporary fixes
rather than coming to me*

Who created thee

*I am here for you to answer every question that
goes through your mind*

*Sometimes I just want you to sit back and
relax your mind and unwind*

Reflect on how I have kept you this day

How I have been there for you every day

I have the answers that you seek

Come and sit and sup with me

I'm knocking at the door of your heart

Don't be distant are stay far apart

I hear you and I will answer your call

Jesus

Therefore, if anyone is in Christ, the new creation has come: The old has gone, the new is here. (2 Corinthians 5:17)

Jesus to know you is life

Jesus I cannot live without your ever-present touch

Jesus you open my heart where it has been broken apart

Jesus your Word washes my mind and keeps me aligned

You are my Anchor

You are my Solid Rock

You are my Shield

You are my Majestic One

You are my Father

You are my Friend

To you Jesus I want to stay connected until my end...

Bleeding Heart

Delight thyself also in the LORD; and he shall give thee the desires of thine heart.

(Psalm 37:4)

You came running to bandage up all my wounds

You allowed me to lay at your feet when I experienced so much defeat

You showed me who I was and gave me new claim to my identity

I have never had to pretend as I stand in your presence

You washed me cleaned and got rid of all the resin

My bleeding heart found it's cure

When I bowed my head in the presence of my master and kneeled at His feet

You have comforted me beyond the words that I could possibly articulate

I am healing from the inside out

Growing stronger by grabbing onto your hand

You lead me out of the darkness and into the light

My bleeding heart has found your illuminating light

Chapter 7
A New Beginning

Great Delight

Who being the brightness of his glory, and the express image of his person, and upholding all things by the word of his power, when he had by himself purged our sins, sat down on the right hand of the Majesty on high. (Hebrews 1:3)

You are wonderous within my sight

Your name brings such a great delight

Nothing else can satiate my desert soul

My eyes lock in and are fixed for life

The Time is Right

***I press toward the mark for th prize of the high
calling of God in Christ Jesus.
(Philippians 3:14)***

The time is now

Go ahead and march forward toward the victory ahead

Relinquish all the defeating thoughts floating around in your head

I am the one that speaks all things into existence

Speak life and you shall have what you speak

In all your speaking make sure it aligns up with me

I will lead you along the path that you must take

Pivot when I tell you which turn you must make

Make your life count and don't waste your time

Time is precious and you do not get a refill

Take this time to fulfill all that is destined in my will

Your purpose and destiny are beckoning you to come forth

Step forth for the time is now to conquer all

Captured

You have searched me, LORD, and you know me. You know when I sit and when I rise; you perceive my thoughts from afar.
(Psalm 139:102)

You have captured my estranged heart and I don't want to know another

Your love is so amazing it leaves me in wonder

I've been captured by the beauty of your character that leaves me awestruck every time I meditate on you

How is it that you can do all that you do

I stand in wonder and left amazed

I keep trying to maintain my gaze

My soul feels with joy to call you Father

Captured by your love that flourishes my soul

Thank you for maintaining the Master's hold

Never letting me go but holding me ever so tight

Daddy Please don't let go of my hand

I'm willing to surrender and walk with you hand in hand

Limitless

The Lord thy God in the midst of thee is mighty, he will save, he will rejoice over thee with joy. (Zephaniah 3:17)

You have caused me to rise above the storm

You called me to yourself beyond the constraints that I and others have placed upon my being

You tell me I'm far greater than I have believed

For you have told me my identity is in you alone

I'm limitless and free

I'm created to Be

I'm the Me You Created Me To Be

Touch my mind, soul, and spirit

To come into agreement with all that you have said

Shape my world to come out of agreement

For all that you have not said

Soaring above the boundaries that have been laid out for me

I'm flying high being the me

You called me to be

Rise Above

***I can do all things through Christ which strengtheneth me .
(Philippians 4:13)***

The Tide Rises and Falls
We have been created to rise above it all
All the challenges that come our way
We have been built different in a special and unique way

We are taught how to lean and depend
Trusting in His Word that will never end
Hanging on to the Hope that lies within

Sharing it with this world that will someday come to an end
We are not exempt from life challenges you see
For the Holy One has given us the strength
To put our trust and dependence in thee

Rising above it all
Does not mean you will not encounter any falls along the way
It just reveals the true discovery of
Who is in control of it all

Instructions

Trust in the LORD with all thine heart; and lean not unto thine own understanding.

(Proverbs 3:5)

What was the last thing I told you to do!
Though the road seems tumultuous along your way
Stay the course and do not deviate from the way

Yes, another pathway may be what you like
But you cannot see all that I have in my sight
Trust the process as you navigate the terrain
Understanding that it will be encompassed with
Snow, Sunshine, and Rain

I have equipped you with the tools to reach your destination
Refer to the instructions in the book that will never change
You'll reach your destination
And receive the ultimate crown

You'll learn in the end why it was never meant for you to go the other way

Chapter 8
Time To Soar

Extraordinary

O house of Israel, cannot I do with you as this potter? Saith the LORD. Behold, as the clay is in the potter's hand, so are ye in mine hand, O house of Israel. (Jeremiah 18:6)

Your life in its totality is
Nothing short of extraordinaire
People cannot help but stop and stare
The beautiful smile on your face
Shining bright for all to embrace
You've climbed through ashes and so much rejection and disgrace
Yet and still
You have remained resilient beyond all the daggers and distaste
For you knew one day you would have a story to tell
To empower, encourage, and coach someone else through
For you discovered the BIG PICTURE it was never just about you

Extraordinary by far, yes you are indeed
You desire for all to win and succeed
You want them to know just how gifted they are within
For them to relinquish control to the Savior
That will bind up all their wounds
You know all to well the anger and rage you carried inside

That's why your heart is compassionate for the one that has been cast aside

Comfort them, pray with them, and let them know they are not alone
They can come to the Father no matter how far they are from home.
Their lives or a treasure and extraordinary too
Share your story that was meant for my glory

Guided By The Light

Then spake Jesus again unto them, saying I am the light of the world: he that followeth me shall not walk in darkness, but shall have the light of life. (John 8:12)

I run into you because you are safe, and your light is so bright

There is nothing that can keep me from your sight

Even when I have tried to run and get away

You always have caught up with me and let me know that you are here to stay

You never changed your mind about me along the way

You light my path and I keep following you your way

Your light is bright

Shining so big so everyone else can see

It's hard to believe you still have time for everyone and not just me

Shine Bright your light that keeps me connected in

I have found it to be my solace that never ends

I'm resting in your presence that has and eternal end

Thy light is hope when I can't understand my way

Thy light is hope that brings me home for another day

Spiritual Formation

But all things that are reproved are made manifest by the light: for whatsoever doth make manifest is light. (Ephesians 5:13)

Growing In You has caused me to give up what I usually want to do
To live in alignment with what pleases you brings peace to my soul
To grow in community causes me to grow in my maturation process
My character continues to grow and develop in the Christlikeness
That you have called me to be

Help me to confront my soul with your word that examines me
Pouring out myself to you is so freeing and delightful in your sight
You have collected every tear and known have
been aborted in your sight

Being stretched to grow is never easy at all
The tearing of the Spirit and Flesh is in a battle to win
Help me to lay hold of you and sow to my Spirit for the definite win

It's not about trying to measure up and do everything right

*It's about my relationship with you that helps
me to lose the dead things*

And chase after you to go all in

For entanglement with the world is not my friend

You're calling me out into the deep where my soul desires to be

My soul has answered the call to where you want me to be

Rest In Me

Take my yoke upon you and learn from me, for I am gentle and humble in heart, and you will find rest for your souls.
(Matthew 11:29)

Many times, you have scurried around from place to place

Trying to outlive the weight of your past

Nothing Is wasted with me

I will use it all for my gain

Can you trust me as you ride the wave

It will be one that will cause you to come unglued and remain resilient and brave

I have called you to come and rest in me

To trust me as you walk through your macroeconomic life

Knowing that true joy is found only in me

The promise is that you will be at peace and free

I teach you how to process through every stage of your life

I want you to know that you will win at life

Find your security only in me

Accomplish all that I have given you

Just don't lose sight of me

There are many more coming after that you will have to coach and mentor along the way

Hold them accountable but walk with them along the steps of the way

Let them know where you found solace and who is your greatest friend

Point them to me

So they to will find rest in me

Triumph

We will triumph In your salvation. In the name of our God, we will set up our banners. May the LORD grant all your requests. (Psalm 20:5)

I am he the triumphant King

Everyone that comes to me triumphs in me

Many do not know me, and I desire for them to experience true lift

Take time to invest in them day by day

Sharing the love of Christ in the beautiful but small ways

Make them feel special for no reason at all

Tell them how your Daddy Pa Pa

Did this for you each and every day

Especially when you were willing to walk away

Love them like the have never known

Call them out in prayer for them to return home

I am their triumphant King

Awaiting with loving arms

To welcome them in

There past and sin is no match for my love

I am the omnipotent one

Citations

This book was written because of intimate conversations with the Father. The purpose of this book is to inspire, uplift and encourage others by letting them know that they are never alone, that their life matters, and you have been created and designed by the Master Creator with a purpose. There is purpose embedded in every human life that walks the earth, and the earth is groaning for them to stand up and be accounted for. The Father has called you to Himself and He awaits patiently for your return to Him to know Him and make Him Known.

Life does not stop there because He has work for you to do that only you can do. It does not matter what your station of life is just know there is a gift inside of you that is ready to be revealed and discovered.

Are you ready to take the plunge? Dive deep into developing an intimate relationship with Master Potter, who heals every wound, gives you beauty for ashes, and rewrites your story. He is waiting! Are you ready to soar beyond what life has been to become all that you have been created to be? The Time Is Now! Get Ready!

The Time Is Now....

References

The Holy Bible: New International Version. (2011). *The Holy Bible (NIV)* https://www.biblegateway.com/versions/New-International-Version-NIV-Bible/

The Holy Bible: King James Version. (1769/2017). *The Holy Bible (NIV)* https://www.biblegateway.com/versions/King-James-Version-KJV-Bible/

The Holy Bible: New King James Version. (1982). *The Holy Bible (NIV)* https://www.biblegateway.com/versions/New-King-James-Version-NKJV-Bible/

The Holy Bible: English Standard Version. (2001). *The Holy Bible (NIV)* https://www.bible.com/versions/59-esv-english-standard-version